Six-Word Lessons for

DATA-DRIVEN
DECISION-MAKING

100 Lessons Today's Data Pros Must Adopt for Exceptional Bottom-Line Results

Daniel Rubiolo

Published by Pacelli Publishing
Bellevue, Washington

SIX
~WORD
LESSONS

Six-Word Lessons for Data-Driven Decision-Making

Published by Pacelli Publishing
9905 Lake Washington Blvd. NE, #D-103
Bellevue, Washington 98004
PacelliPublishing.com

Cover and interior designed by Pacelli Publishing
Cover image by Pixabay.com
Author photo by Alpha 1 Photo & Studio, Bellevue, WA

ISBN-10: 1-933750-81-2
ISBN-13: 978-1-933750-81-1

Dedication

To all of you with a passion for bringing positive change to the world, considering learning as a never-ending endeavor.

And to those who fight against the luring of procrastination, energizing everyone around you with your bias for action and personal growth.

Contents

Acknowledgements

Sometimes people may perceive others during debates as not having strong opinions, just because most of the time they focus on listening and comprehending first. That kind of describes me. Circumstances permitting, I like to understand the perspective of others during discussions, ensuring I do not miss a chance to learn a better stance before sharing mine. Devising Information Solutions require a great deal of this in order to drive clarity and consensus across many people and groups with conflicting interests, as to reach the most valuable common ground and priorities. I fully respect and admire people with strong skills in organization, planning, structure, focus on driving quality results, low tolerance for laziness, recognize the potential from people and demand it, and do all of this with a deep loving passion. I dedicate this book to the one person who embodies all these, and whom I admire the most: my wife, Marylin. We like to joke about how well we complement each other, balancing empathy with action, and ideas with planning.

I consider my two sons, Federico and Francisco, as the two other people who motivate me to perform at my best, because they inspire me every day with their commitment to excel and their thoughtfulness in approaching challenges as they keep growing into young adults. Among many other things, we share a passion for sports, through soccer (Leo Messi's Barcelona) and martial arts ("Choi Lay Fut" style of Kung Fu).

As it happens with almost everything, we all need a strong foundation, like growing up influenced by examples of organization, attention to detail, quality, empathy, planning for the long term, dedication, and passion to achieve significant results. My parents, Hugo and Graciela, exemplified all these principles for me and my brother, Javier, leading us with solid morals too.

Finally, a recognition to my coworkers over the years. I always valued people striving to improve themselves (learners), who fight against procrastination. As well as those with passion in what they do, and how they do it (quality and pride in their craftsmanship). And those that help grow all others around them, contributing to their results and coaching them. Especially all the ones who tried to keep up with my wondering mind and ideas: I know it required lots of patience at times. And mostly, the ones that pushed back, forcing me to improve.

Even the ones who thought adulation would take them to a special standing or would influence my perspectives; they never understood what motivates me (in short, they did not get it, and still don't); yet, I found lessons in their behaviors.

And my gratitude to: Jim DuBois for encouraging me to write this book; Manisha Arora (Patel), Martha Laguna, Rebecca Seerveld, Srinivas Kanamarlapudi, and Diego Baccino for proofreading and plenty of feedback; Michael King for writing the foreword; and Gregory Weber for our daily insightful conversations about our experiences on applying industry best practices.

I also mention throughout the book, in order of appearance: Hernán Pueyo; David García; Marissa Meyer; Jose Gasser; Hugo Monella; Alejandro Lagares; Carlos Cortes; Diego Caracciolo; Armando Calva; Rodolfo Mendoza; J.D. Meier; Martin Lee; Serguei Gundorov; Geoffrey Sears; Marcelo Weber; Karina Derbez; Kiki Tsagkaraki; Carlos and Karina Mainero; Gustavo and Flavia Zbrun; Marc Reguera; Dario Bonamino; Eva Medran; Rick Stover; and Hugo Petrucci.

Thank you all!

Foreword

As an IT professional for more than thirty years I have seen a lot of decisions made and a wide range of data maturity at the point those decisions were made. Effective leaders who have to make decisions without data take some common additional steps. They make their decisions time bound, clarify scope, communicate what the decision does NOT mean, and create room in their decision communications to iterate the decision when they get new data. For me this lesson book is all about ways to mature the last point. One of the first things I push for when I'm working with someone on data analytics is to try to put that person into the shoes of the leader who has to make the decision. In many cases that leader will have to decide regardless of the quality of the data delivered, and getting the data-provider for any project as close to the leader's decision-making requirement as possible has repeatedly accelerated the delivery of value inside a major project. I had the honor of working closely with Daniel throughout most of the learnings he describes in this book and was constantly surprised by the agility he brought to our strategies. The ability to change approach, integrate new data, and take a stand when the data said we were off-course were all things that set Daniel's work apart.

Much of this book talks about the early stages of the journey to iterating on your maturity models and the pitfalls you can avoid by paying attention to the drivers of pressure outside of the data itself. While I highly recommend nailing the early recommendations here, I also believe the nirvana of any strategic investment is achieving the end goal. Doing that requires the ability to focus on both the tactical churn and the end state concurrently. I've done multiple three-year projects leveraging Daniel's concepts in this book, some with the advantage of Daniel executing the analytics side of the projects. On all these projects the early investment in analytics didn't "feel" like it was paying off. Yet on each of these there were "moments" where the light went on for all the leadership consuming our status and the vision looked achievable. Each of these "moments" came because we had iterated our data, and were willing to totally change which concept we used based on both the new learnings in the project and sometimes new feedback from leaders. In one situation specifically, we were struggling to get our stakeholders to take action that we knew inherently would make their IT systems more stable. In partnership with others, Daniel delivered the reporting possible to predict future failures to an accuracy of seven days and we were able to use the data he provided to change a reactive conversation with business leaders into a proactive one. The end result was that action on newly discovered defects became as important to our stakeholders as responding to actual downtime events.

The reward for anyone passionate about data maturity is seeing that data in use in a way that increased decision quality.

Michael King
Service Engineering Group Manager
Microsoft

Preface

The turning point in my career into applying BI techniques to aid decision-making started with a fortuitous conversation. At the time, I worked on data center consolidation projects, leading the standard operational procedures and measuring progress through data collection and reporting. I saw the potential to scale up this automation and serve the entire organization with end-to-end data processing, modeling, presentation, and use of high-quality information to improve business discussions.

Armed with my STP presentation ("Situation, Target, Proposal": a concise way to organize a proposal for discussion, evaluation, and approval), I made my case with the general manager of my immediate team. However, even though he ultimately agreed with its value for the organization, it did not align with his team's priorities, and he could not approve the resources at that time.

That afternoon we had a team event, and I happened to stand beside another general manager, Jim DuBois (who later became Microsoft's CIO):

- Jim: Hey, how are things?
- Me: Good, very good. We continue to make great progress toward the consolidation. But we could do so much more by automating our metrics, and we cannot at the moment...
- Jim: Tell me more...

- Me: (I explained my proposal to him, going through my STP again, but this time without the slides.)
- Jim: I love it! I will sponsor it for the unit. Talk with your manager again, and if he agrees, let me know, and we will proceed.

That started my professional career in using data automation to support business decision-making. Since that day, the learning never stopped. Not just about technology (which evolves fast, opening new opportunities all the time), but also about industry standard best practices in designing these solutions, formatting data for presentation to executives, storytelling, program management, organizational dynamics, and much more.

From my experiences since then, I chose the 100 key lessons I believe will bring the most value to your projects on designing business processes and developing information solutions to make data-driven decisions. Of course, given the available space in this book, I cannot cover all the details but will point you in the right direction. In the accompanying blog, http://rubiolo.net, I will share more information, including extended discussions on definitions, implementation, and technology.

I hope these lessons will give you some ideas so that you too can build your path to these experiences, and devise ways to capitalize on them. I feel committed to your success, so please take advantage of the resources at my blog, including the contact form for questions.

Enjoy!

Introduction

Tackling the complexity in devising solutions to provide factual information to aid decision-making drives to very rewarding learning!

These types of solutions (commonly called "BI Solutions") seem unique because many fields of business and technical expertise must get together and work in coordination. They produce business metrics from data across business processes, directly correlating to high priority business strategies. Executives with various levels of experience and expertise use them in meetings and business reviews, usually involving varying degrees of ambiguity, and where many different situations get discussed all the time, such as new and continually evolving hypothesis and ideas on how to proceed.

These solutions do not stay still--they evolve all the time.

The broad meaning of Business Intelligence involves a never-ending, dynamic process of continuous learning. It starts with business users needing information (some type of intelligence). As the solution begins providing reports that executives can use for more informed decisions, many new questions and hypotheses on potential courses for action will arise, requiring more information. And there the cycle starts again and will continue to evolve by constantly adapting to change.

Users changing their minds on requirements will seem not only common, but also expected as they evolve through this learning process.

To achieve success, this cycle requires not only technology in managing data and information systems, but also many processes on driving human behaviors (the "culture"), project management, team dynamics, and much more. Take advantage of the many industry-standard best practices available to organize, plan, design, develop, deploy, release, operate, and improve a successful solution.

The meaning of "data-driven business decision-making solutions" implies two contexts in play. First, a technology data processing solution, designed and implemented as a Data-processing Information System with BI reports and analytics automated for business consumption. Second, encasing the first, a business context, in which these users apply the information provided to their business reviews in order to assess the health of the business, potential options, and make decisions on actions to take. One concerns itself with technology, the other with people processes.

The following chart describes the recommended framework to use in these projects, as a guiding model to compel maturity into business processes and assist in maximizing results. It focuses on using proven statistical techniques in designing BI, with the goal of driving deep understanding on:

1. How to model business processes, and what to measure about them.

2. Identify the root causes leading to underperformance, and drive quality.
3. Given current restrictions, assess the best potential results one could achieve at the time, and identify gaps.
4. Evaluate the possible actions one might take on the gaps at this time and choose the best one that would maximize profitability.
5. Add the ability to run what-if scenarios and forecasting models to enable more advanced business scenarios ("AI" as in "Artificial Intelligence" goes here, too).

Business Process Maturity Framework

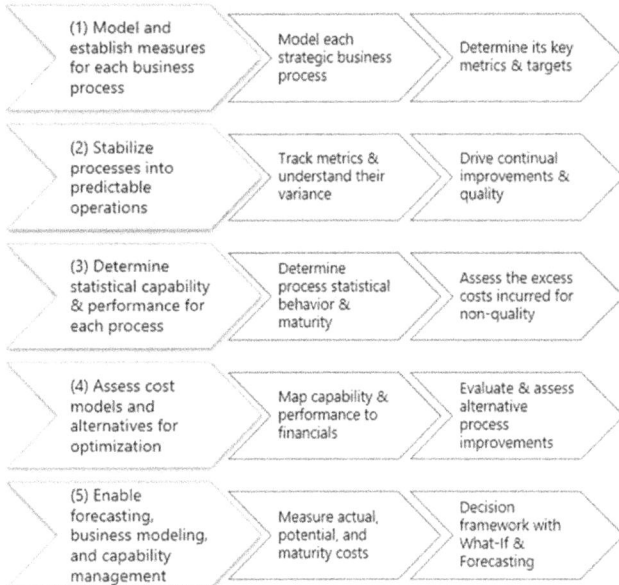

(1) Model and establish measures for each business process	Model each strategic business process	Determine its key metrics & targets
(2) Stabilize processes into predictable operations	Track metrics & understand their variance	Drive continual improvements & quality
(3) Determine statistical capability & performance for each process	Determine process statistical behavior & maturity	Assess the excess costs incurred for non-quality
(4) Assess cost models and alternatives for optimization	Map capability & performance to financials	Evaluate & assess alternative process improvements
(5) Enable forecasting, business modeling, and capability management	Measure actual, potential, and maturity costs	Decision framework with What-If & Forecasting

Along with the business leads and domain experts, model the business strategies to measure (using techniques such as Balanced Scorecard) and design the business processes (with techniques such as Strategy Maps and process models). These generate all the leading requirements for the outcomes needed from the Information System.

The following picture describes the recommended architecture framework to structure and design these solutions:

Data-processing Information System Design Pattern

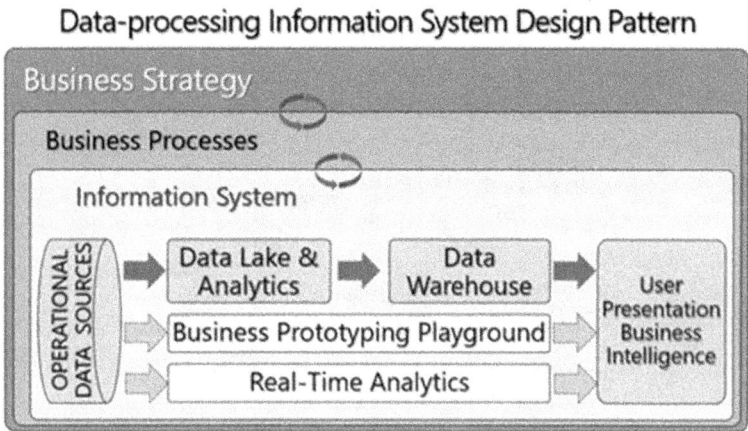

The remainder of the book explains these models, with Part I and Part II describing them, and Part III detailing how both complement each other.

PART I:
PROCESS MATURITY

Chapter One
People Make Mistakes When Making Decisions

Projects to build decision support systems require teams with a blend of acumen (both business and technical) and a fervent dedication to learn, recognize mistakes, and correct course.

I can trace my experiences on this topic back to two of my initial managers, Hernán Pueyo and David Garcia. Hernán led a fast-growing subsidiary, and he drove innovations on several product lines, including investing in the comprehensive information systems to manage all this, for which he hired me to design and build. David managed IT Research for the branch of a multinational company, in which I drove technology standards, and the quality initiative leading to ISO certification. They had expertise in their business domains, and eagerness to learn more; they would make decisions to move ahead directionally to drive progress, but ready to adapt and correct course as we learned more. I feel proud of having them both as friends. Our conversations always inspired us to carefully lay out our opinions, understand opposing views, and concede to the best argument. We bounce ideas off each other, learn, and grow together.

As Montier[1] articulates, humans usually learn early how to make easy choices fast while postponing important decisions for later, making humans "predictably irrational." Emotions and fast decision-making served the species well to survive, but in a modern world, they lead to mistakes, like chasing momentum without fully evaluating the ramifications. To improve decision-making everyone must understand this natural bias and learn how to make better "rational" choices.

In order to process vast amounts of information rapidly, human brains evolved to make quick judgments based on approximations rather than precision. Logic requires effort and time. Given the fast pace of life today, most often everyone uses emotional reactions to make decisions, and rarely analyze them logically. From a behavioral psychology standpoint, "by design, emotion will always trump logic" (Montier[1]). Using logic too often depletes energy, so humans cannot use it constantly.

Consequently, devising a plan when not under pressure, and then committing to it, stands as a great technique to minimize emotions getting in the way of decisions in the heat of the moment. To do this effectively while driving business decisions, consider the following as crucial action steps:

1. Define strategy ahead of time.
2. Carefully craft KPIs (Key Performance Indicators: the intended business outcomes to achieve).
3. Set deadlines for expected results (up the pressure to increase the focus).
4. Gather data to measure progress towards achieving them.
5. Automate as much as possible the analytics between the goals and what the data tells.

Now, the key lessons on this topic...

1

Set deadlines to get things done.

If the need for information comes up during business discussions, commit to a deadline to set leadership's expectations on related decisions. At least commit to a date by when to complete an evaluation of the effort required for the answer. Setting deadlines imposes a sense of ownership and urgency, which drives focus and avoids procrastination. However, use with caution: if everything demands attention, then nothing does.

2

Define a strategy guiding toward results.

Everyone needs some "northstar" (guiding strategy) to focus efforts to get something done. Strategies provide for this, bringing everyone together with a common goal to attain. They may point to medium or long-term objectives, evolving as people learn. They may structure in layers, with longer-term ones driving short-term goals. When deciding how to measure them, set targets defining milestones toward success for the current stage. Measures without strategy have no purpose.

3

Drive thoughtfulness by setting appropriate targets.

Targets drive action, guided by strategies. They may reflect year-end or monthly goals, but should always have a time constraint. If people push too hard, resources get strained, usually negatively, impacting other areas of the business. Resources (people and investments--the who and what) constrain not only what people can do, but also business processes design (how). When setting targets, consider who will do what and how, by when.

4

Automate analytics assessing gaps to targets.

Reaching a target means that the team appropriately planned the "who does what and how by when," given the guiding priorities and strategies. Conduct analytics to identify constraints on missed targets. Maybe the team needs more people, time, investments, or a different process. Focusing analytics on the causes for missing the previously planned targets will inform future planning, options and course corrections. This defines learning.

5

Use data to drive business reviews.

Data processed with context transforms into information, which can drive action through decision-making. Consider information as a precious asset because of this. Everyone should consistently have access to this asset (data) to help move the business forward. Making information widespread across the organization better equips a culture of using data to make better decisions to attain business goals. Executives must not only use data for informed decisions, but do so openly to drive the culture by example.

6

No data? Decide directionally; revisit later.

Sometimes the business cannot measure something they know they need to know. The process to collect the data might not exist yet. Avoid freezing until all data becomes available. Show leadership by deciding on a hypothesis and pursuing it. Later, demonstrate openness by adapting to the learning. For significant implications, reduce uncertainty by first running experiments to validate the hypothesis. Not deciding also counts as a decision. (See DuBois[2] on building the mindset to accelerate change.)

7

Focus on learning, not proven right.

Never focus on getting data to justify a decision. Instead, get data to learn and assess next steps toward continuous improvement. "Final decisions" only feed egos, not business results, as they prevent adapting to better options and drive the culture to a negative spiral preventing better results. A learning culture reduces biases and opens up to new opportunities. (Search for "fixed mindset versus growth mindset" or look at Dweck's book[3].)

8

Don't blindly follow authority; bring value.

In a growth mindset culture focused on learning, executives do not expect people to blindly follow decisions that might lead to the wrong outcomes. They expect the teams to use data as part of performing actions to assess impact, and if someone learns that the decision might lead to problems, they expect the team to speak up and avoid those negative consequences. They don't want anyone justifying later with, "I was told." They expect everybody to proactively bring value and challenge the status quo.

Chapter Two

First Things First: Modeling the Business

Quoting Kaplan & Norton[4]: "IMAGINE ENTERING THE COCKPIT of a modern jet airplane and seeing only a single instrument there." That does not convey much confidence, does it? Yes, one needs only a "few" key indicators of performance to focus efforts on delivering the most critical business results, but sufficient ones to understand and fine-tune as one progresses.

As a member of the CIO Scorecard Team, I learned about this from my manager, Marissa Meyer. She relentlessly led us to apply Six Sigma quality practices to ensure discovering and focusing on what drives the real priority for the intended business results, including "measuring even when we couldn't" and navigating complicated politics towards common goals (assessing competing priorities, driving consensus, and defining shared objectives).

Attain clearly defined metrics and carefully planned goals by using statistical thinking, principles, and processes. These remove most of the uncertainty (mostly in interpretation) while steering the thinking in making evident and easier to navigate the complexity in data, the business processes that generate it, and how to handle it more adequately. (More on quality and statistics applied to business scorecards: Gupta[5], Balestracci[6].)

In addition to tracking progress in attaining desired goals, metrics also guide the organization toward common, clear outcomes. The team can organize them in layers. One metric may represent an entire area, while people in it will manage many more at their level. Alternatively, the same metrics may cascade to the individual areas by tailoring the targets specifically to them. In any case, the team must make the mechanisms of relating/cascading very clear and accessible to everyone involved.

Other books by Kaplan & Co.[7,8,9] and Niven[10] provide guidance on defining and organizing Key Performance Indicators (KPIs) tracking strategies, organizing them into scorecards, and visualizing everything with Strategy Maps. This framework provides a comprehensive approach to understanding the strategy, translating it into plans of action, and as a communication tool for aligning the entire organization.

Now, the key lessons on this topic...

9

Carefully craft all metrics and KPIs.

Strategies should reflect in the hypothesis of what to do with current resources, investment levels, and market constraints. Break them down into metrics that will fluctuate with the actions across the organization. When they have defined targets, they turn into "key performance indicators" (KPIs). Additional metrics may exist, but a select few will drive the essential priorities for everyone, forcing focus and alignment on execution, and leading to the realization of the strategies.

10

Don't use averages for setting targets.

Often, managers plan targets with "the average consumption last period," not realizing that it means that roughly half the time it will underperform, driving excess costs. Lacking statistical thinking to consider the system or process view that generates the numbers, these simplified "hacks" will miss targets too often, and potentially significantly. Everyone involved must understand the concept of "statistical variance" and the causes for the highs and lows, in order to plan appropriate and comprehensive targets.

11

Model metrics as belonging to processes.

Every measure reflects an element of a business process, driven by constrains from its design, resources (people, investments), capacity, etc. Everyone must understand the processes producing these numbers, assess their inputs (volumes, types, velocity, trends), their consumption (categories, quality, costs, utilization, capabilities), and their outputs (quantities, satisfaction, quality, velocity, ROI, costs). This results in a comprehensive view of the business, and the definition of the KPIs may evolve with learning and maturity.

12

Look at distributions when setting targets.

For setting targets, look at the process behavior that produced the metric. The statistical distribution of their values will reflect everything needed to understand the process and decide how far to go. Everyone must understand the variation in the numbers, in order to explain cases of normal behavior for the process with its current constraints, versus unexpected ones that need further investigation. Only then will the team feel empowered to take appropriate actions.

13

Plan for seasonality when setting targets.

Initially, management may set a desired year-end target for a KPI. The team will investigate and analyze possibilities, potential action plans, and even data collection. With more information available, they can evolve to monthly targets, which reflect maturity in understanding the process and possible actions. They may require increments or reductions. And they will adapt to the seasons (analyzing seasonality in the processes). This process leads to realistic targets.

14

Balanced Scorecard: the comprehensive business view.

As with all models and frameworks, Balanced Scorecard and Strategy Maps provide invaluable guidance on what to consider and how to structure when modeling strategy and their measures for the business. They also apply as powerful tools to communicate priorities and guide the organization into aligning efforts, plans and actions toward common, shared goals. After building the information system, they assist in organizing the presentation of the KPIs into effective scorecards and cascading dashboards.

15

Manage differences between scorecards and dashboards.

Scorecards organize KPIs into quick overviews of the overall health of the business, driving attention to critical areas. Dashboards provide the next level of detail, mixing KPIs with other various related metrics and charts (and even text) to assist with fully understanding the area of interest, managing possible actions/decisions, or providing more context and explanations. Usually, scorecards cascade into dashboards, and they may relate to other dashboards and operational reports.

16

Build maps from a defined strategy.

Consider a picture worth a thousand words. To drive common understanding and focus throughout the organization, linking the KPIs with the strategy into Strategy Maps represents a great way to communicate progress. Use these maps as a tool for business reviews and communications. Also use them to structure the automation of Business Intelligence, linking Strategy Maps to scorecards, to dashboards, to projects and plans, and to operational reports.

17

Display metrics even without any data.

Sometimes a specific topic requires metrics that at the time cannot result from known or available measurement systems. Avoid getting frustrated. Do not ignore these metrics. Add them to the plans, KPI lists, and communications anyway. If leaders drive a learning culture nurturing a growth mindset, the organization will come up with and offer ideas on how to manage them. Showing clear intent to accomplish something, even without knowing how, will foster creativity across the team.

18

Measure it to accelerate getting results.

If priorities do not have definitions (strategy, KPIs), action plans do not have specifications (projects, targets), teams have not received clear and extensive communications (scorecards, maps, dashboards), and all of it does not have effective measures (statistical thinking, data collection and modeling, information system, reporting), then it will not get done. Leaders must structure plans, drive a learning culture biased for action, and measure progress to make sound business decisions.

Chapter Three
Understanding How Far One Can Go

A chieving an information system that provides the measures feeding decision-making defines the start of business understanding and potential for real impact. Initially, this information will only tell how the business processes work. Apply statistical thinking into interpreting these data in order to listen and understand the processes that produce it and come up with options to intervene and improve them.

My dear friends Jose Gasser and Hugo Monella taught me about the value of looking through the obvious and into the fundamental ideals; to avoid distractions from the essential by ignoring the superfluous; to appreciate and cherish the wonder in the simple things and its effect on thoroughly enjoying life. Many of these lessons apply to business modeling as well.

The pervasive statistical illiteracy observed today grows by the day into a threat to business success. Understanding what variation in data means, and its adverse effects when misunderstood, turns out as critical skills for decision-making in today's fast-paced, competitive, and complex business environments.

Think of the journey to data-driven business decision-making across the following maturity phases:

1. Gather data to understand the business processes and how they run today, enabling operational decisions toward achieving targets.
2. Use additional data and statistical tools to assess process stability and put management processes in place to handle improvements and deviations (understanding variation and special causes of unpredictability).
3. Additional data and analysis will allow to fully understand the potential capacity of processes (given its current design and resources), enabling the evaluation of the gaps and missed opportunities, as well as assessing in detail the various options for improvements.
4. With more data, start estimating the current operational costs incurred, as well as alternative courses of action, and evaluate the ones that will lead to maximized profitability.

5. All these data will enable automating powerful what-if analytics and more precise forecasts.

It defines a journey of iterative data collection, analysis, learning, and decisions. Data enables driving continuous business improvement, as well as to assess the need for, and impact of, more significant changes (when incremental enhancements turn up insufficient). An information system, a growth mindset and learning culture, along with statistical thinking, becomes the secret formula for effective decision-making.

Works from Balestracci[6], Wheeler[11,12,13,14], Savage[15], Campbell[16], Huff[17], and Few[18] provide detailed definitions and guidance on these areas.

Now, the key lessons on this topic...

19

Get prepared to orchestrate data analysis.

Well-known industry standards describe how to understand, interpret, and communicate the message contained within the data before taking appropriate actions on the underlying processes. Collecting data for proper analyses involves much more than just building data warehouses, reports, and scorecards. It requires learning about the processes generating the numbers, their contextual meaning, and knowing what kinds of decisions executives need to make. Statistics provide appropriate tools for this.

20

Statistical thinking: understand and minimize mistakes.

Anyone can compare two numbers, (e.g., this month versus last month). Hint: one will measure higher, and nobody would know more than that with any truthfulness. Real progress derives from understanding how the team obtained and interpreted them, what they should do with them, and in which context. Today's industry best practices on using statistical methods for decision-making provide the tools. Take advantage of them. They will assist in understanding what the numbers try to communicate.

21

Avoid significant misinterpretations by understanding variation.

Nothing stays exactly the same all the time. Everything evolves and fluctuates. Teams must fully understand variation as a fundamental notion, and how it relates to KPIs, to correctly interpret numbers. They cannot take action on common-cause variation in the presence of special cause variation. Prevent wasting time and resources on red herrings. Understand the concept of statistical variation to avoid bad decisions.

22

Handle variances, not points in time.

Business processes produce numbers for KPIs over time, and variation will always show up in them. Interpret these numbers in context and as a trend, and don't just compare them as a couple of points in time (e.g., monthly, or year over year). Without considering the history and trend between the two points, nobody can sufficiently explain the difference; only that one number measures different than the other. Again, the team must understand the notion of statistical variance.

23

Adjust targets as learning evolves.

In a growth mindset culture, don't obsess with "just" hitting the target. If teams ignore the lessons from the process, they will drive a culture of blaming instead, always looking for excuses. Use these lessons to help executives consider adjusting targets as a healthier option. Better results will derive from adapting the targets than hitting the wrong outcomes (particularly when the available information indicated this, but everybody chose to ignore it).

24

Drive results' predictability by stabilizing processes.

The design and resources available to run processes dictate its outcomes (reflected in the KPIs). When analyzing variation, finding instability indicates the presence of special-cause variation. This means the team can investigate and remove these causes, driving to stability, until they can detect nothing more than the common-cause variation inherent to the processes' design and resources. These stable processes produce predictable outcomes, resulting in better chances to reach business expectations.

25

Variation always drives business process behavior.

At the risk of "beating a dead horse," unmanaged variation causes business underperformance and cost overruns. It really does. Variation always drives process behavior. Managers who do not understand variation can keep screaming at people demanding improvements, but the fear this intimidation produces will only drive more unexpected, random changes (more variation), possibly resulting in even worse outcomes. Again, learn about variation. (End of the beating.)

26

Assess business process behavior with statistics.

A tool called Process Behavior Chart (PBC) assists in handling variation. As a particular case of statistics' control charts (X-MR chart), a PBC graphs the values measured from the KPIs, interpreted in context with its statistical boundaries (in turn based on practical mathematical models). The chart tells when the process becomes unpredictable because a special cause exists, which now the team can investigate and eradicate. It also assesses if a known change had the expected impact to results or not (experiments). Use it.

Ideally, business processes will show predictability.

PBCs will also tell when processes show predictability, and what to expect from them in the future. Consider a process predictable when no observable special-cause variation affects it: the process runs as designed. No significant fluctuation in the numbers exists, as long as they fit within the statistically expected ranges. These ranges represent the statistically predicted behavior for the process. However, it might not align with desired targets. Make decisions to affect the process and manage the gap, if any exist.

28

Drive predictability to create more consistency.

When the PBC shows that the process attains statistical predictability, the KPIs will reveal consistent outcomes within the parameters of the process design and resources. To improve it, additional changes should follow. It could manifest as the need for more resources, or a different procedure, or something else. The PBC will indicate whether those changes moved the process in the desired direction or not. In this context, it behaves as an assessment tool.

29

Entropy trashes the business process behavior.

The PBC drives processes toward predictability, and then it assesses: (1) if something new happened that affected the process (or if it continues to affect it) and needs investigation (detection mechanism); or (2) whether or not changes resulted as expected (validation mechanism). Both enable decision-making. If the team does not use PBCs, over time the natural course of things (entropy) will degrade and undermine the processes. Keep using PBCs as an alerting tool.

30

Blame unpredictability on pesky assignable causes.

When the PBC indicates unpredictability on a process, it means it has unmanaged "assignable causes" affecting its results. Armed with this knowledge, the team now can investigate these causes, identify them, and act to remove them. However, if the PBC indicates the process shows statistical predictability, nobody can explain a difference between two numbers as if a reason existed. The process works as designed in that case; no excuses.

Chapter Four
Many Options - Which Path to Take?

Once processes display stability and predictability, start evaluating options to improve them. Making a process stable does not mean it reached optimal levels for what the strategy requires. Move into assessing why the process capability differs from what one expects it to achieve (targets).

I have four friends in particular who illustrate for me that a combination of acumen (knowing the business), openness to learning new ways, and passion for pursuing better options always drive to positive business impact: Alejandro Lagares, Carlos Cortes, Diego Caracciolo, and Armando Calva. Each of them works in different business areas but share traits that make them very effective in driving progress forward. I always felt very inspired by them.

With today's technologies it turns out easier and more affordable to base analytics on all data, reducing the need for sampling (except for accelerating performance during exploratory investigations). This makes analytics much simpler to apply and interpret. It also provides traceability when enabling the organization to "drill-down to details" and see all the cases that led into the conclusions, and thus make root-cause investigations much easier, valuable, and widespread across the organization. This enables everyone to learn, get informed, and act to improve results.

However, this learning still needs guardrails to protect everyone from random actions across the organization that will produce more variation. Many times, improving one area may degrade the results of the overall process. This happens when an improvement on one part of the process moves the bottleneck to a different part, where it becomes more expensive to handle or makes the process break due to lack of capacity. All new actions, even when well-intentioned, need an assessment of impact, analysis of critical path and bottlenecks, as well as an overall understanding of how they will affect results.

In addition to the works mentioned earlier, other books from Wheeler[19,20,21] provide excellent guidance on this area.

Now, the key lessons on this topic...

31

Categorize data: experimental versus observational studies.

When collecting data to run experiments (experimental studies), expect signals in that data to show the impact of the intervention to the controlled factors included in the experiment. When collecting data from a running process (observational studies) you do not know what to expect. The techniques to interpret data in these studies have significant differences. Experiments will assess the potential impact of an intervention, and then with observational studies, confirm the real effect on the process.

32

Understand improvement types: precision versus accuracy.

The PBC tells what the current process can do. If it shows predictability, then it also indicates what to expect from it in the future. If that expectation differs from desired results (targets), then make changes that may improve accuracy (values closer to targets) or precision (most values closer together). The combination of these two options produce four possible types of changes. The PBC will reveal which one to pursue as next step.

33

Use statistics to direct process maturity.

Many easily available analysis tools enable identifying assignable causes, even simple ones like Fish diagrams, Pareto charts, and other root-cause and problem-solving techniques. However, nothing beats talking with the experts – the people running the processes. Analysts or managers never know everything; they must continuously learn and investigate. Most of the time they will perform as facilitators for the experts, working together, iteratively, with data, to drive business process maturity and predictability.

34

Decide on a specific statistical methodology.

Lean, Six Sigma, SPC (Statistical Process Control), or some other? Do not get stuck with just one way. The resources (people, training, materials) and products, services, processes will determine the most appropriate. These methodologies have some primary focus (e.g., waste, process) but they largely use the same tools, and complement very nicely. Usually, processes will have so much unpredictability, that simply understanding variation, using PBCs, and some root-cause tools will bring immense value.

35

Define what quality means to everyone.

Citing Wheeler[20], processes achieve quality when, "We design processes for delivering on target, operated predictably, and with continual improvements in place to minimize variation" (which in turn continually reduces its costs). This means operating them toward full potential. They get better and better over time at delivering to expectations while maximizing value. The traditional definition of quality as merely delivering to target does not imply optimized function. The team can do much better by pursuing full potential.

36

Assess costs of nonquality and nonconformance.

A quality process does not have unpredictability. The disrupting effects of having assignable causes require additional cost to manage them. Also, there may be extra costs when the process (even if predictable) does not conform to targets (e.g., reprocess, rebuilds, scraps, inefficiencies). Model all these costs and make them explicit to management. Many times, the best improvement comes from reducing waste, duplications, extra steps, bureaucracy, and other excess costs first.

37

Predict the future with process capability.

To paraphrase Wheeler[20]: Data reflects historical events. All data analysis relates to historical examination. Yet all the questions of interest pertain to the future. So how do you bridge this gap? Welcome to "process capability," which characterizes what a process will produce in the future. For this, the process must show predictability already. For unpredictable processes, one can only presume "potential capability," while working on making it predictable by removing assignable causes.

38

Assess maturity by calculating process performance.

For any process, one can assess its performance based on past data (its performance index). For a predictable process, it equates to its capability (its capability index). However, if it shows unpredictability, then nobody can use the past to predict. It becomes a measure of hypothetical capability, only achieved if driven to predictability. These indexes, along with the PBC, help decide on types of improvements, and enables tracking progress toward better results.

39

Estimate process potential by calculating capability.

When analyzing capability, also consider how it relates to the targets or specifications. How far do they stand from each other? This will result in two additional indexes: centered capability index and centered performance index. These four now really allow assessing the current situation and the best types of decisions to drive improvements. Basically, how can good executives consider the process and what it produces?

40

Assess the excess cost of operations.

With the indexes, the team can estimate the relative "costs of operation and use" and assess the expenses incurred because of the gaps between them (as a consequence of how much off-centeredness and unpredictability the processes show). Explicitly exposing the excess costs incurred when operating inefficient processes allows assessment of options for improvement, their own costs, and the ranking of them. This enables objective, data-driven decision-making.

41

Understand the options for process improvement.

The team may face up to five situations, each of which implies specific options for courses of action. The first four come from a combination of how much stability the process shows, and how on-target results it produces (the outcomes represented by the KPIs). The worst, a chaotic situation: off-target with an unpredictable process; then, on-target but still unpredictable; next, off-target predictable; and finally, the ideal: on-target predictable. The fifth presents the attainment of an optimized cost.

42

Endure chaos: off-target results, unstable process

Situation #1: When the process shows unpredictability, the most significant gains will come from stabilizing it, by getting to understand variation and using the PBC to drive the investigations into assignable causes, and then work on removing them. If the outcomes also show non-compliance (off target), additional improvements should also focus on centering to targets (driving accuracy). This requires the most interventions to bring all four indexes together.

43

Center results, even with unstable process.

Situation #2: In this scenario, the performance indexes indicate the current situation, and the capacity indexes show what the team could hypothetically get out of the process if stabilized. Focus on driving results closer to targets (centering), by managing the centered indexes (centered performance index and centered hypothetical capacity index; this last one because the process shows instability), taking actions to bring them closer together.

44

Drive process stability when missing targets.

Situation #3: Another course of action for unstable processes focuses on using the PBC to identify the existence of assignable causes behind the instability, investigate them, and drive activities to get them removed. Keep doing this until the PBC indicates the process enters predictability. In this scenario, focus on the capacity indexes by driving predictability (moving them out of hypothetical).

45

Achieve the ideal: centered stable process.

Situation #4: Using the PBC to drive processes to predictability (making them stable) brings all the performance indexes closer together. Because the process now shows stability, the performance equates to its capacity. And because it delivers to targets, the centered indexes now align. In this ideal state, the process predictably delivers on-target. From now on, keep using the PBC as an alerting tool, to ensure staying in this ideal state.

46

Even better: let's now optimize cost.

Situation #5: If the processes reflect centeredness and stability, they already reached an ideal state. All investments and resources deliver results as desired, and with reasonable expectations of continuing to obtain those results. To achieve even more, minimize costs via process optimizations by increasingly reducing variation, ensuring plenty of capacity in the process. Remember the definition of quality: predictable centered processes with minimal variation.

47

Model and evaluate all the options.

Why do all these steps help with better decision-making? Because the team can now fully assess how the processes run, what they produce, in which state they perform, and what alternative actions to take on each to optimize them. At this point, use the strategic business priorities to evaluate where to drive additional investments and efforts, to further achieve anticipated goals. This provides a practical model for business decision-making.

48

Establish an early detection alert system.

Natural entropy will always try to push processes back into unpredictability. When using PBCs, the team makes a powerful learning tool available to everyone in the organization. Keep using them as an alert tool for discovering new assignable causes affecting a process, so the team investigates and maintains process predictability. It also assists in assessing if planned changes to the process affect them as expected, or not, and by how much.

PART II:
TECHNOLOGY

Chapter Five

The Modern Information System's Master Architecture

So far, the lessons have provided a comprehensive sense of the data needed, and how to use it. What about ways to collect and organize it to produce the information the users require? How about techniques and best practices for data lakes, structured data, and data warehouses? The following chapters cover these, in light of the model mentioned in the introduction.

Building these systems require a particular type of artistry. My uncle, Rodolfo "Rudy" Mendoza, personifies the role model on this topic for me. He developed a perfected skill in craftsmanship, becoming an example of how patience, dedication, and expertise lead to high-quality results (with also a high dose of intolerance for laziness). Everyone needs these traits when building information systems.

The overall architecture incorporates all aspects of data processing, starting with capturing the raw data from the sources that produce it, to how users consume and utilize it; going through quality assurance, integrations, enhancements, keeping history, modeling for ease-of-use, optimized for performance, and security. It requires many roles with various skills (both technical and business), including heightened aptitude on attention to detail, quality, and customer empathy.

Consider as the most important takeaway that "data has no value until structured for consumption."

Users employ data from reports to assess situations and options to make decisions. Well-designed data models will provide the flexibility to answer a high variety of business questions, drill into details, aggregations, filtering, ordering, relationships, trends, etc., without requiring additional reprocessing as users' needs for information evolve. This has a focus on high availability, concurrency, performance, security, and ease of consumption. Techniques like data warehousing and semantic models structure data in this manner to serve end-user reporting.

However, raw source data usually does not have these qualities. In many cases, their sheer size may make it very difficult, costly, or even impossible to feed reporting directly. Besides, the team can find immense value in preparing these data by adding quality, integration, consolidation (or disaggregation, as needed), extensions, enrichment, and cleaning, before exposing to users. Pre-processing techniques assist with this, organizing and landing data in systems such as data lakes for keeping history. Data scientists then go through these looking for value, and when they find it, data gets structured for consumption by becoming the curated source for feeding the data warehouse. These steps convert "potential" value (data lake) to "consumable" value (data warehouse).

Consider the following data processing areas, derived from proven best practices in designing pipelines:

1. Landing: periodic raw updated data gets saved here directly from source systems.
2. Staging: data quality and business transformations get applied to the raw data, increasing their value.
3. Loading: the curated data gets prepared for loading into star schemas / dimensional models (e.g., ensuring referential integrity, manage early and late arriving facts, mapping surrogate keys, etc.).
4. Schema: historical data gets accumulated from daily loads and stored for consumption.
5. Reporting: views of the schema for consumption (e.g., isolation of internal changes, security, etc.).

Organizations can take advantage of decades of accumulated knowledge about industry standard best practices in architecting and designing for these scenarios (see Corr[22], Kimball[23,24,25,26]).

Now, the key lessons on this topic...

49

Analytical approaches: from top or bottom?

Data in the lake enables a bottom-up approach to analytics (inductive), in which data scientists explore the raw data trying to find patterns to form a hypothesis of process behavior (data produced by some business process). Data warehouses enable a top-down approach to analytics (deductive), in which business analysts have a hypothesis that they look to confirm by navigating to more details. In any information system solution, organizations need both to fully enable business decisions.

50

Data has no value until structured.

The "potential" value of the data in the lake comes from having it available for advanced data science analytics and further processing (such as Artificial Intelligence, Machine Learning, etc.). These usually produce a structured result, which the team can then feed into the data warehouse or reporting layer for user consumption. The "actual" value of the data gets realized when a business user can apply it to making a decision, not before.

51

Correlation will never replace causation's insights.

Some advocates of so-called "big data" propose that accumulating all possible data and letting AI and ML algorithms find significant correlations will drive to a day when understanding of causation will become obsolete. This type of thinking leads to immense risk. Making decisions without understanding why can drive to all the wrong outcomes. (They say similar things about data quality, domain expertise, and models. See Few[27] for details.)

52

Strive for clear and measurable goals.

Ideally, information systems producing insights would focus on "clear and measurable" business goals, with resources focused only in those areas. However, that implies a "solid, mature business process," which usually does not correlate with reality, driving the need for insights. They should also focus on business potential, providing enough information to help the business processes reach higher maturity, while outlining "clear and measurable" business goals.

53

Establish data contracts with the sources.

Teams managing operational systems usually do not know how users will consume the business process' data they produce, so they may make changes as they see fit. But as decisions start depending on these data, ensuring stability and quality at source becomes critical, leading to the need for establishing data contracts by agreeing on service levels and support procedures. These should also include shared goals and joint evaluations of progress in order to ensure mutual success.

54

Sometimes the source doesn't exist yet.

Business users may apply some categorization to the data that does not exist in any tool or system connected to current pipelines (they may even relate to testing new business ideas by, for example, grouping products into ad hoc categories). The team must fully cover these very valid business scenarios. For supporting these, building small data acquisition front-ends connected into data processing will effectively create new valuable data sources.

55

Land only the needed essential data.

Store source data in its full fidelity within the "landing" area, which immediately makes it available for early raw analytics; a "snapshot" of a given period (e.g., a day's worth of updates). Then process it through the automation pipelines. This pattern intends to minimize impact to sources by obtaining only the least amount of data and processing needed. This acquisition activity might also rarely perform some data quality and simple curation, in order to ensure traceability for the data flowing from the sources.

56

Stage preprocessed, curated, high-quality data.

Add value to raw data when transferring it into "staging" by performing quality checks, handling errors, closing data gaps, ensuring referential integrity and operational validations (e.g., stop processing if data flow changed more than normal, and raise alerts). "Business process analyzers" (packages of transformation business logic) make it easier to maintain the pipeline (plus aggregations, disaggregations, classifications, etc.). This pre-processed staging becomes the curated, quality source for further processing.

57

Better together: data lake + data warehouse

The combination of data lakes and data warehouse techniques provide an end-to-end architecture enabling most decision-making scenarios. They deliver the most significant potential for value, taking advantage of data types like unstructured, semi-structured, and structured, while optimizing for each. For example: What happened and why? What keeps happening? Which relationships exist? What will happen? What if? How much risk? What should happen? Which options? How to optimize? Who gets impacted and how?

Chapter Six
Data Lakes Enabling Advanced Data Analytics

Recent advances in technology for data processing and storage focus on modern-day challenges brought up by the sheer volume, velocity, and variety of corporate data, which gets produced at overwhelming rates. Some companies have coined the term "big data" to package these into new offerings, as if these techniques presented something new rather than the continual technological evolution the market has enjoyed for decades (see Few[27]). Don't consider any particular tool as the only one for getting a job done; instead, continue establishing agility and pragmatism when developing solutions for different problems.

Looking to find, adapt, and embrace the best ways to drive effectiveness in all we do evolves into a fantastic skillset, much needed in cases like this. Nobody I know reflects these as well as my friend J.D. Meier. He has developed a mastery in bringing efficiency and purpose to the right problems in the most successful ways.

Today's modern systems (such as: event processing; real-time data analytics; semi-structured data analytics; real-time streaming; network analysis; search; social media; text processing; etc.) have benefited from advances in distributed computing processing and scalable distributed file systems, optimized to take advantage of low-cost, scale-out processing and storage options. Notably, "data lakes" have evolved to handle massive, diverse data volume and have progressed to increase speed, mostly supporting data scientists' scenarios like advanced analytics, machine learning, and streaming.

However, data lakes do not make the concepts and practices of data warehousing obsolete. No data has value until structured for end-user consumption. Data lakes follow a design engineered for data exploration, discovery, learning more about the data, text processing, innovation, flexibility, and scaling out for data scientists who know how to write their questions. They employ less mature technologies, involving more risks. Contrary to popular belief, they often don't come cheap, because they require lots of expensive engineering labor, and advanced features still need licensing.

If appropriately unlocked, data accumulated in data lakes can provide valuable information. But until then, it only offers "potential for value": data scientists search for the patterns in the data to unlock that potential, and then the team structures it into a data warehouse with the purpose of exposing that value to users for applying in decision-making. The design techniques and purpose of data lakes and data warehouses display differences as well as similarities. When architected together for complementing their strengths, they produce better results.

Now, the key lessons on this topic...

58

Optimize some scenarios with data lakes.

Data lake technology follows a schema-on-read design approach, in which data gets stored, parsed, reformatted, and cleansed at runtime when needed, through batch jobs processed with high parallelism. Its engineering design strives to hold a vast amount of data in its native format, using file architectures supporting any type of data (structured, unstructured, semi-structured, raw; e,g., JSON/XML APIs, web-logs, click-streams, social networks streams, sensor and machine data, and even video and audio).

59

Apply data lakes to appropriate scenarios.

Currently, data lake processing technologies lack critical concepts behind mature data warehousing, such as: limited ability to enforce referential integrity or match key lookups; high-performance features like indexing, concurrency and aggregations; metadata management, data lineage, quality and profiling, robust error handling and fine-grained security; data governance; data quality and stewardship; and conformed mastered data. However, they slowly continue to catch up. Find the right scenarios.

60

Data lake: the glorified file folder

Data lakes organize files with multiple formats in folders on top of massive parallel storage using low-cost hardware. Their design focuses on staging and preparation of data, batch processing, historical data storage, and as a sandbox for data scientist's exploration. However, they do not provide optimizations for recurrent, reusable end-user reporting and queries, security and compliance, low latency, interactive ad hoc queries, or high concurrency. They excel as a source of clean data for data warehouses.

61

Data governance prevents lakes becoming swamps.

Data lakes follow optimizations for high availability, scalable, distributed file storage for raw formats of any type of data. Paired with batch processing, streaming, social media, and video/audio, etc., it can quickly grow in size, number of files, complex folder structures, and more. They need further maturity in supporting data governance to ensure these data meet enterprise standards, business rules, data definition, data integrity, lifecycle, and stringent security, risk, regulatory, and legal controls.

62

Avoid "data lake versus data warehouse."

The data lake ecosystem continues to evolve, while data warehouses have reached maturity and continue to grow supporting scale. Avoid thinking in terms of "one or the other," because they have different design goals and strengths. The fundamental principle asserting that "not all data is created equal" implies that one must choose the data storage and access mechanism to best suit the usage of the data. Think exploration (data lakes) followed by consumption (data warehouse).

Chapter Seven

Designing and Building Modern Data Warehouses

Common practice defines a data warehouse as a hierarchical, structured data repository (following a schema-on-write pattern) of integrated/consolidated data from multiple disparate sources, organized and curated (providing governance, security, and reliability), with the purpose of creating mission-critical analytical business reports for end-user consumption in decision-making.

For end users to adopt this layer of the overall information system in making business decisions, build them with a multidiscipline and high performant team, both technical and business. On these topics, I learned with and from great professionals like Srinivas Kanamarlapudi, Martin Lee, Serguei Gundorov, and Geoffrey Sears. They live to and inspire high standards in areas like: the value of in-depth technical knowledge; profoundly caring for clarity and quality; relentless customer empathy; as well as expecting and demanding the absolute best from solution providers. Everyone needs all these qualities when engineering these data solutions.

Traditionally, a data warehouse manifests itself as a central repository of structured historical data on top of a relational engine optimized for performance, reliability, availability, and ease-of-use. Complementing the data warehouse, usually a "semantic" model, adds an optimization layer to make it easier for client tools to expose more advanced features that facilitate end-user reporting and ad-hoc querying (pre-aggregations, relationships, advanced calculated measures, hierarchies, row-level security, and more).

However, consider its two principal components separately: the design principles and best practices, and its technological implementation. Because data warehouses typically run on relational databases, sometimes the two get used interchangeably, leading to many gruesome misconceptions when discussions like "data lakes versus data warehousing" arise. Its design patterns, best practices, and business goals remain as critical today as ever.

Data warehouses manage current and historical business process data, by correctly modeling the accumulation of data changes over time to represent the transactional business events meticulously. Raw data gets processed for things like quality, master data management, integration, and cleansing, as part of the ingestion pipelines.

The data warehouse delivers the centralization of the "historical single version of the truth" needed for decision-making across the enterprise; the single point of reference for structured, reliable, believable, comprehensive, accessible to everyone, high-quality data. Their design focuses on high concurrency (many simultaneous users) and high performance, and it follows "dimensional" modeling techniques that bring business events and entity language into facilitating end-user understanding and consumption.

Now, the key lessons on this topic...

63

Model star schemas, dimensions and facts.

The Kimball approach to "dimensional modeling" has established itself as a widely industry-accepted best practice in designing agile data warehouses. It structures the design into "star schemas" containing: (1) "dimensions" that represent business entities your users will apply for navigating, filtering, aggregating, and organizing data in reports; and (2) "facts" as a very specific design pattern for detailed transaction information optimized for high performance, serving as the source for all business metrics and KPIs.

64

Drive consumption and navigation with dimensions.

Dimensions represent business entities by aggregating data into denormalized tables (e.g., customer, products, services, offers) with all related details in columns (e.g., names, address, categories). Each row/record gets assigned a unique warehouse key (numeric surrogate key), which allows mixing related data from different systems, keeping the history of changes, and maximizing performance. Users pick fields from these dimensions to aggregate, filter, and organize metrics about business events into reports.

65

Measure business process events with facts.

Facts measure business events or transactions (e.g., utilization of devices, sales of products, occurrences of an incident). When these occur, a fact table relates together various dimensions (e.g., a specific customer bought a quantity of product at a price on a particular date from a salesperson at a location). A "fact record" groups all the related dimensions' surrogate keys and involved amounts/quantities (e.g., the count of product, unit price). This maximizes scale, query performance, and calculations.

66

Provide metrics and KPIs with facts.

Because fact records mostly contain numeric fields, the system can highly compress them, enabling high performance and scalability when answering business questions (e.g., total sales by location, inventory of products, best sellers, trends over time). They also optimize performance for query engines to relate facts with their dimensions through joins, as well as to calculate complex logic for aggregations, hierarchies, metrics/measures, and KPIs (in both relational and semantic engines, using the same model).

67

Agile development with the bus matrix

The enterprise bus matrix evolved as a great tool as part of Kimball's data warehouse bus architecture, which has rows for each business process event (facts) and columns for each dimension participating. It defines the scope of work (all facts and dimensions needed) and interdependencies between them, and the team can use it as an agile planning tool by feeding the backlog with work items to build. It also proves very effective as a communication tool to track progress.

68

Increase business value by extending staging.

Sometimes the team does not have a source for specific types of data, such as classifications, categories, or hierarchies. They need to provide a home for these business components, usually with a lightweight UI to allow users maintain these data, and then integrate into staging to enhance and augment the raw feeds. These tailor-made extensions to functionality provide one of the most significant value-adds for business reporting.

69

Strive for smart and intelligent staging.

Produce data warehouse-worthy data in staging with: smart extraction; data staleness checks; splitting or combining raw data feeds; detecting errors and gaps, auto-fixing where possible; alerting and auto-ticketing; stopping pipelines if the source presents detectable corruption; tracking traceability of data moving through pipelines; implementing referential integrity; preparing for historical recording by adding surrogate keys; pre-calculating aggregations; and transforming natural keys into dimension keys.

70

Model business entities, not just reports.

The information system's goal of mimicking the structure of the business processes allows users to understand what happens and make decisions accordingly. Sometimes users will ask for specific reports they need short-term, but one must see beyond that. When modeling business events (with facts and dimensions at the lowest grain), that would already cover the reports they wanted (and many others). By modeling business events, one enables a plethora of insights, anticipating the most common questions.

71

Design solutions to enable the future.

Do not design for the reports available today; instead, build the capability to unlock the business future. Using the information system, users will get metrics from all business processes and will start asking more in-depth questions, thus drilling down to more details (data leads to understanding, which leads to needing more data). A well designed dimensional model delivers high scalability with modern technologies, especially with the advent of the Cloud and the continuous reduction of costs.

72

Keep the business processes' lowest "grain."

The "grain" defines the level of detail for data collection on business events (facts). In a dimensional model, it represents the number of dimensions by which one can analyze those facts. Sometimes, in the name of saving space, someone may consider reducing the grain by aggregating the data. Do not do this if possible, because it will sacrifice drill-down ability from users, limiting how much they can investigate their data.

73

Architect data warehouses: hub and spokes

The data warehouse represents the "hub": the central high-volume processing unit and repository of curated, conformed, high-quality, efficiently designed historical data. It pushes its data to "spokes": (1) semantic models that optimize for end-users' ease-of-use consumption, advanced calculations, query performance, high-concurrency, and security; and (2) data marts optimized for automated systems. This pattern also enables high availability, as the spokes remain accessible while the hub runs its processing.

74

Perform processing with ETL data pipelines.

Data pipelines perform all orchestration and processing required for ETL (Extract, Transform, Load). They effectively implement the logic of getting data from sources into landing, kicking off data processing tasks in staging, orchestrating coordination of these tasks, triggering alerts and notifications, keeping historical records of these activities, etc. They consume the most time of the overall information solution development, because they deliver all the action that needs to happen.

Chapter Eight

Learning to Fail Fast: Playground Prototyping

As mentioned earlier, information systems assist in managing a business as a never-ending process of testing evolving hypotheses through decision-making. In many cases, these hypotheses do not live long once they served their purpose (such as validating the applicability of an idea), so the team may either discard them quickly or develop them further until landing on some useful learning that users can consume.

An exploration activity such as those for analytics requires a curious mind, with an intrinsic motivation to find what the data tries to say while avoiding the pitfalls of letting preconceived notions distort the facts in the data. Natural human tendencies make us take shortcuts to simplify the complexity around us, thus adding lots of bias (as explained in the initial chapters), so analysts get trained to manage against these tendencies and unlock the filters people build over time in how they perceive things. I learned (and keep learning) a ton of valuable lessons on this topic from my friends Marcelo Weber, Manisha Arora (Patel), Karina "AniKa" Derbez, and Martha Laguna. They taught me lessons such as how our different thinking styles complement each other and allow us to understand events from different perspectives to solve problems, or brainstorm different approaches to an issue. This exploration, prototyping, and challenge to our perceptions help us grow and gets us closer to realizing our full potential. I consider them very inspiring.

Sometimes, users need as quickly as possible the raw data from analysts and data scientists to maximize value in their research. The Business Modeling Playground in the solution provides a place where data entities go through a much faster lifecycle, prioritizing flexibility and speed in support of immature, evolving business processes. These experiments, once validated, will provide requirements for their full integration in the overall information model.

In particular, the results of data scientists' analytic work would indicate what needs to get added to the ETL (data processing to Extract, Transform, and Load) into the data warehouse for user consumption, or discarded when no longer useful. Implement this either as part of the data lake or the data warehouse, depending on how fast it needs to get exposed to users, their level of technology skills, and requirements around privacy, security, or data governance.

Now, the key lessons on this topic...

75

Deliver business value with rapid prototyping.

Show openness to assist users in accelerating their learning. Most of the time, in order to develop something, one must have well-defined requirements. However, equally often the business does not really know what they need. They will have questions about the value or feasibility of specific data in an area, so making it available to them with minimal processing from the team's part will help them determine the potential value of it and clarify requirements.

76

Ensure continuous support for learning culture.

Very likely, business users will have data they generate that the team would need to integrate into the solution temporarily to evaluate alternative analyses of business impact. Some client tools might help make this easy in some cases, but a more flexible solution would enable a "playground area" in the information system where analysts can save custom data and use it to evaluate these options.

77

Conduct experiments to deliver more learning.

Another common scenario requires enabling some immature data streams from experiments to flow through the information system very flexibly. These streams may relate to new apps or workflows under testing by the business, which users need to correlate with existing data and determine the value for a process. These experiments may live for a short period, and change frequently, so the team must enable a dynamic data flow process in the solution to support this important business scenario.

Chapter Nine

Integrating Real-Time Feeds into the Solution

Real-time data feeds usually do not belong to information systems for a couple of reasons: (1) these feeds commonly support some specific business process for operational management, not higher-level business decisions, and (2) they bring real challenges in curating and integrating them fast enough to keep their real-time characteristics. However, in many cases, incorporating them as part of the overall framework of the information system produces significant business value.

Quickly taking advantage of opportunities enables the business to grow, to reduce time-to-market, or to realize the starting of a new trend and jump in to "ride the wave." The same applies when analysts bring a business idea to management's attention, and they get support to pursue that idea and attain its value (or "fail fast" by realizing quickly that maybe it does not apply). Two leaders in particular I had the pleasure to work with and learn from, Michael "Mike" King and Jim DuBois, demonstrated many times this vital skill of sensing and nurturing the best in people and supporting their initiative in testing the limits of a process in search for business value, while developing the talent potential in their teams.

Typically, real-time feeds get routed into dashboards for front-line employees, who use them to make tactical decisions while executing a business process. Normally, consider these real-time feeds as part of operational applications, and not the information system. However, in a few notable cases, some real-time feeds may need modeling alongside the information system to inform management of a particular and remarkable series of events. For these cases, set up a specific part of the solution to handle them.

Data warehouse techniques do not support real-time needs, because they rely on curating and modeling the data in efficient ways for high concurrency querying and reporting, and these real-time feeds cannot wait for that processing to happen. Data lake techniques also do not support them, given their optimization for file batch processing, also adding delays. However, several new technologies have recently emerged to better handle these scenarios by extending data lake or data warehousing methods, such as connecting directly to dashboards visuals, beefing up reports with real-time trends.

Incorporating real-time data as part of the information system instead of line-of-business applications results in a common reason for taking advantage of technology consolidation (keeping similar technologies together) and reusing skillsets from dev/ops employees.

Now, the key lessons on this topic...

78

Real-time feeds highlight business trends.

Information systems usually do not integrate real-time feeds because of their use in specific business review meetings, decisions, or assessments (usually time bounded, not continuously), in which the curation and integration of data provide the most value. However, sometimes having real-time trends alongside the rest of the business information may help in complementing some particular decisions. In these cases, the data supports activity closer to operational processes rather than executive ones.

79

Expose business events as they happen.

Another typical case for real-time feeds relates to informing as soon as possible when specific events happen. They act a bit more closely like alerts, rather than counting the events (and thus reflecting some type of trend). They usually relate to rare events that may need critical actions, such as reacting as fast as possible, triggering investigations, or consuming significant resources, etc. This enables operational management of these situations, and also informs executives of key developments as they occur.

Chapter Ten

Delivering Insights: User Presentation Business Intelligence

This part of the information system makes available for consumption to users everything built so far. This layer comprises a set of tools in a "user presentation" area, which includes automated reports, alerting, ad hoc query tools, presentations, dashboards, portals, and more. Use these during business discussions to aid in decision-making. Most of the user training, support, and utilization will occur in this layer.

The team needs a very particular blend of skills to outline and enable this layer of the information system. My friend Kiki Tsagkaraki personifies a great example of a professional always championing for users while building these types of components. She demonstrated over the years a blend of customer empathy, business acumen, and a relentless focus to empower business users by providing them with comprehensive analytics driving understanding in making decisions.

As builder of the information system, provide some structured and recommended ways to access the data, including flexibility and comprehensiveness, so that users can take advantage of it based on their skills. Unlock the information and make it available for business use (under the appropriate security policies). If one restricts access or methods of consumption, it would significantly hamper the potential business value and ROI from the information system built.

The builder role does not in any way allow anyone to own and act as gatekeeper of the information generated. It enables business value and decision-making without playing a role in deciding who, when, and how will access and consume the data. Business process owners must own access control. As technology expert and enabler, recommend procedures, security controls, standards, and toolsets for more effective use (also methods for support, documentation, monitoring, data accuracy, and training). Facilitate utilization of the system in the most efficient ways possible, to drive continuous learning of the business through the soundest decisions.

In this context, getting to "insights" means delivering appropriate information at the right time. This simple definition implies the on-time delivery of the curated data from across business processes into ROB cycles (Rhythm of the Business, i.e., business reviews, projects checkpoints), making it available for answering high-value ad hoc business questions, and presenting it in ways appropriate to its optimized use (e.g., alerts, notifications, email, reports, scorecards).

Now, the key lessons on this topic...

80

Provide Data as a Service (DaaS).

Imagine the information solution becoming the "go to" place for the business information that users need. They have confidence in its accuracy. They know the team delivers it always up-to-date. They have seen the team continuously adding more value all the time. They like the reports provided, finding them effective and insightful. They rely on its information to make the most important decisions. They use it to plan, check, verify, decide, and monitor the business. Data runs the business by its use in decision-making.

81

Empower business users with semantic models.

Information ready for decision-making requires source data appropriately modeled in a way that makes it easily accessible and highly optimized for end-user consumption and ad hoc querying. A semantic model includes well-defined business data entities, their relationships, hierarchies, business KPI logic, change history, dictionary definitions, and security in a comprehensive repository. Their optimizations include high concurrency (simultaneous querying), high performance, report authoring, and ad hoc analytics.

82

Data marts within user presentation layer.

Expose data mart(s) to advanced users, data scientists, and developers across the organization to help them take advantage of curated data in their automation processes. Their implementation usually manifests as relational databases through appropriately secured SQL APIs (e.g., tables, views, stored procedures) or as file dumps in a lake (in commonly used formats), which deliver widely available options to development, analytics, and scripting tools, and follows optimizations for high volume throughput and querying.

83

Create persona-based data access patterns.

A diverse user base includes executives, data scientists, managers, business analysts, and operational personnel, etc. In making options available for their effective use of the data, the solution will provide dashboards, scorecards, canned reports, real-time feeds, ad hoc querying, experimentation, data mining, exploration, and mobile consumption. These require different designs, implementations, tools, and training, etc., combining into reasonably complex requirements for specific data access patterns.

84

Avoid taking shortcuts for short gains.

Just because throwing unoptimized designs to a tech may work for today's workload, do not ever take this shortcut because it will not scale and will soon increase costs for required fixes. The patterns shared earlier derive from mature industry learning and proven design standards. Plan to unlock all data and optimize it for consumption (performance, scalability, availability; having low-level grain to enable in-depth analytics with drill-down to details; etc.).

PART III:
SETTING IT ALL IN
MOTION

Chapter Eleven

Avoid Misleading People When Presenting Data

Anyone may unintentionally "lie with statistics" and data very easily. As explained in the initial chapter, the human mind evolved to reduce complexity into quick shortcuts, which includes limitations in how the eye interprets things like shapes, areas, colors, and patterns. Show caution in how to present data and charts, so they do not mislead or confuse the audiences, while also simplifying for them the understanding, interpretation, and effort in "getting the message" contained in the data.

I relate my inspiration on this topic to the trustworthiness found in friendship. Case in point, after about twenty-five years of hiatus I met again with my friends Carlos and Karina Mainero and Gustavo "Tavi" and Flavia Zbrun, and it felt like time had not passed. We still enjoyed sharing the wonder in the simplest of things, the transparency and openness in our conversations, the clear stories and messages, and the connections we formed. When designing for presenting data to others, follow those same qualities leading to trustworthiness.

Design reports intentionally. Which context produced those numbers? What does the chart try to say? How do these numbers relate to the business process they represent? How do those charts relate to each other? In which order should someone consider them? Consider these and more questions when designing reports, charts, and other types of data presentation and visualization.

Also, visualize the business process underlying the data, as it will define the right course for interpreting the data correctly. Design the "story the data tells" in a way that correlates with how the process generated the data, providing the full context in which it happened (the inclusion of other related data).

Also, explicitly state the type of interpretation supported. Does someone need it for exploration? Or for concluding an analytical procedure? Or to indicate current standing in achieving targets? Or to assess where risks lie ahead? All of these require excellent skills in storytelling.

And then, consider the human mind limitations mentioned earlier. Some people cannot distinguish all colors (color blindness). Most cannot precisely assess volumes from areas in shapes. Others find it difficult to extract trends from a tabular list of numbers. All of these and many others require adapting the data presentation designs to account for these limitations.

To learn more on information display design principles see: Few[28,29], Knaflic[30], Robbins[31], Forsha[32].

Now, the key lessons on this topic...

85

People cannot stomach pies and doughnuts.

Some chart types prove challenging for the human mind to interpret correctly, leading to confusion at best, and totally misleading at worst. Charts that depend on assessing areas of shapes (e.g., pies and doughnuts), while easy for computers to generate, prove difficult for the mind to judge. Other charts may combine colors that some people cannot distinguish, or their labels appear hidden, overlapped, or applied to the wrong context. Or they cram too much information together. Avoid them.

86

See to believe: pictures communicate better.

The combination of the human eye and mind supplies a high-speed image processing engine, which the team can take advantage of to communicate better and more effectively. Pictures of data in the form of graphs can pack a lot more information to discern patterns much more quickly, compared to the same information in, say, tabular form. The mind delivers high-speed comparisons for things like size, lengths, enclosure, closeness, connections, and position.

87

Choose visualizations for analysis versus presentation.

During analysis and exploration, analysts will use many different types of charts in slightly different ways, because they try to find something in the data. However, when they find a meaningful insight, they should use different, more specific charts for presenting to users, designed for ease of interpretation and impact.

88

When everything demands attention, nothing does.

In presenting stories with data, highlight where to drive the attention to. But not overdo it, because if the highlights make everything scream for attention, then nothing will. In business review meetings with executives, analysts should start with the critical stories for discussion, then go to the overall Balanced Scorecard (which delivers the story of how the KPIs interrelate with each other, and the progress toward achieving targets).

89

Expose noteworthy situations: measure rare events.

Very often, processes face incidents or situations that need special handling and may consume significant resources. After some time driving improvements, these events get reduced until reaching a point where they rarely happen. Converting these counts into "time between events" or "longevity" allows better measurement of these optimized processes, switching focus to a fine-tuned quality characteristic. And the team can still apply statistics to detect deviations and keep the process stable.

Chapter Twelve

Making Decisions Using the Information System

A fter adopting the recommendations described thus far, executives will ideally express things like:

> "*Every morning I look at the scorecards and drill into the dashboards for some of the KPIs. Usually, everything runs smoothly, but sometimes I find things like a trend that catches **my** attention, or an alert because the system detected a deviation. And in those cases, I immediately have access to additional reports, or I ask the system an ad hoc question, and I get **my** answers most of the time. When I cannot get it, or when I think **we** need more investigation, I have direct*

> *access to the appropriate people and get them engaged. This information system has really empowered all of **us** in managing much better.*"

The scenario requires an unyielding obsession with understanding how to build solutions to empower users. Customer focus, along with a drive for product quality, drives enormous business value. A former manager, Marc Reguera, exhibited these qualities in driving crisp customer scenarios into product planning. Also, a couple of people I have admired for a long time, Dario Bonamino and Eva Medran, taught me valuable lessons on the power of relentlessly aiming for essential goals through focus, dedication, and effort, until achieving our aspirations. These traits allow attaining empathy for the users' struggles with data solutions, and from that viewpoint, devise better designs and options for them to achieve greater efficiency at their jobs.

However, nobody should consider the goal as having everything on target, but to use the information system for making decisions as a never-ending cycle of continuous learning. Every KPI achieving green means the team does not push enough to promote learning. Some areas may require stability for quality reasons, but strategic objectives need learning to encourage growth.

Every business process has some recurrent business review meetings, in which managers and analysts discuss with executives the health of the business. Where the business needs decisions, one should provide an assessment of the situation, along with potential options for action, including a recommendation from the team. The information system would ideally demonstrate the case and illustrate these options, along with what-if analyses indicating the potential impact of each one. Everything discussed so far in this book fits together here into the comprehensive solution that enables business understanding and decision-making.

Now, the key lessons on this topic...

90

Apply maturity matrices to drive adoption.

Build a matrix with business processes in the rows and the maturity phases they will go through, as the team delivers data for them (discussed early in the book). This approach communicates the entire scope, the progress so far, what follows next, sets the right expectations, highlights challenging areas (bottlenecks), and emphasizes which ones already use information to drive discussions. This effectively acts as a roadmap of the solution and business adoption.

91

Establish data contracts with the users.

As the information system approaches completion, make very clear all expectations to business users: How often will the data get refreshed? How will they ask questions about the data? How can they escalate issues they find in the system or data? How fast will support respond? Where can they find the roadmap? How do they ask for more features? Answer all these (and many more) in a data contract for the DaaS.

92

Shape data-driven business review decisions.

The following will accelerate the learning cycle and maximize value: (1) an executive sponsor that bans static presentations during business reviews, encouraging every process lead to onboard into the information system (I learned this from a team leader, Rick Stover, who consistently did just that); and (2) Involve the solution team on these reviews, so they know the context, how people use the data, and allows them to devise additional system improvements.

93

Improve as capabilities develop over time.

How does the system add value and evolve? To answer this question, develop an "information system quality scorecard." It would measure critical indicators such as their ability to get predictable with their performance and capability indexes and the trends of these measures over time. This would demonstrate the continual improvements towards business process maturity due to the use of the information system, complementing the roadmap with a representation of business impact.

94

New shiny things: technology keeps evolving.

Technology develops endlessly in many areas, such as today's major advancements in Artificial Intelligence (AI), Machine Learning (ML), Advanced Analytics, Data Visualization, and IoT, etc. Consider these "data-related" technologies that promise to reveal more business value. However, clearly discern what belongs to the information system versus new feeds that serve a business process. This difference will allow honing focus on producing information to aid decision-making, versus enabling operational feeds.

Chapter Thirteen

Functional Roles: Nothing Happens Without People

The following list enumerates some of the industry standard best practices and technology knowledge needed along the lifecycle of building these information systems: data lakes; statistics; text mining; data mining; relational normalization; data quality; Six Sigma/Lean/SPC; Hub & Spokes; BAM/BPM/ITIL; process modeling; indexing; in-memory column-store; star schema/dimensional modeling; in-memory tabular; SUCCESS; Tufte/Few; DAX/SQL/R; Agile; program management; and quite a few more.

Hardly something any one person may fully know. This makes evident the need for multidisciplinary teams.

All this knowledge and skill requires training and exposing the team to experiences that will allow them to develop. I learned this from my longtime friend Hugo Petrucci, who embodies the qualities needed to ensure people grow as the project progresses. He senses the best in people and nurtures it, so they have opportunities to learn and flourish. By aligning their aspirations with the business needs, he enables them to experience how their efforts, contributions, and opportunities directly benefit the business. The project success becomes a personal success, so everyone wins.

Building the solution will need different roles and knowledge at various stages of the lifecycle; and even others at different phases of maturity in those stages (the continuous learning). This means the team can plan for training its members to bring them in at the level needed, complementing their learning with consultants where they need specific knowledge faster. It turns into a career path, in addition to the product roadmap.

And finally, on the issue of team dynamics and how to work through the entire project consistently and efficiently, Kimball[28] proposes a flexible, agile lifecycle framework that organizes the work of building the information system, which one can easily extend to cover aspects of business process maturity, playground/prototyping, real-time feeds, user training/feedback loops, alignment with business reviews, and ROI/NSAT assessments.

These hint to an essential aspect of building an information system solution: the critical skills needed must include a mix of DBA and MBA. A pure blend of DBA technical roles can design the databases, feeds, processing, and reports, but will fail to deliver the right information users need to drive their business decisions because they will not understand the business dynamics. Likewise, a pure blend of MBA roles may fail to balance technology limitations/constraints and setting the right expectations back to users. Clearly, these projects need a mix of these complementary roles to close the gap between technology and business.

Now, the key lessons on this topic...

95

Build solutions efficiently using Agile approaches.

The variants of matrices discussed so far apply to scope the work, highlight progress, and communicate next steps, etc. All those develop into requirements, feeding a list of work to do, which the team keeps in priority order working with sponsors and the users' community. That prioritized backlog feeds the short dev/ops cycles that build the system, making it really agile to changing/evolving business needs. The Agile approach to development proves very well suited to developing information systems.

96

Understand the software versus data engineering.

These engineering disciplines focus on very different roles, aiming for distinct objectives. Both deal with logically converting a raw input by using some technology to provide a result. However, software engineering primarily builds products, applications, and services from given requirements. Data engineering focuses on acquiring, cleaning, transforming, disambiguating, enhancing, and preparing data feeds into useful formats for analysis. Primarily, the skills of data engineers will deliver the information system.

97

Define critical skills of data engineering.

Data engineers hold responsibility for managing data analytics infrastructure, e.g.: tools and components of data architecture; database solutions; data warehouse design; ETL design, architecture, and development; and a relentless focus on meaning of data, quality, completeness, and lifecycle. This requires high doses of curiosity, problem-solving, and willingness to truly understand the business objectives of the data. These prove as essential as the engineering and computer science skills.

98

Core skill: collaborate well with others.

As discussed, the team needs a multitude of skills, roles, and experiences to build the information system, from profound technical infrastructure, to dev/ops , management, analysis, and business acumen. Meaningful collaboration across all these disciplines proves paramount for success. The typical "rising stars/know-it-all/type A/cowboy/I-can-do-it-alone" personalities instigate a detriment to results. They put morale at risk, hinder synergies, and weaken the value that collaboration brings to overall outcomes.

99

Invest heavily: analytical skills; business acumen.

People represent the key asset: they produce everything that gets done. Nobody would stop priming their crucial asset, would they? To develop people's careers and business outcomes at the same time, invest in improving skills. Focus training on data science skills (e.g., data analytics, statistics, technology aptitude) as well as business acumen (e.g., strategy, metrics, process modeling). Better skills applied with a full understanding of the business will drive to superior results.

100

Succeed in understanding the customer's needs.

Develop empathy to learn the user's job activities. That perspective allows mapping the team's technical knowledge into devising better ways to make those job activities more efficient and successful. Something the user perceives as complex may be easy for the team to develop, such as quickly enabling raw access to a new data source. Or it could represent a very complex endeavor that the user perceives as simple. Understanding these fully assist in helping business users.

Conclusion

Building an information system manifests as using data engineering and technology in the cycle of continuous learning that enables managing a business through decision-making.

In these type of projects, do not build a static solution, but a dynamic, always evolving and adapting system that provides data as a service. It requires agile dev/ops methodologies to coordinate a variety of roles to this common goal.

These projects do not provide some reporting options on top of operational systems, but a consolidated, curated, enhanced, trustworthy, and secure source of information to aid in decision-making. These characteristics of the solution tell users much more about the business than what they already know from operational applications.

These systems produce business metrics from data across business processes. Executives with various degrees of experience and expertise use them in business review meetings. This happens with varying degrees of ambiguity as new and continually evolving hypotheses and ideas get assessed.

These require driving an appropriate culture enabled by a learning cycle in a growth mindset environment. In this context, technology solutions provide data to business processes, which in turn utilize it to assess the health of the business, evaluate potential options, and make decisions on actions to take.

The beginning of this book described a framework to design and drive business process maturity by using industry standard statistical techniques to maximize results. Next, it detailed a second framework for architecting and building the information system: the technical data processing solution that implements the business process framework designed with the first model. They work together synergistically over time, building upon each other as the overall solution matures.

In joint discussions with Jim DuBois[2], we arrived at the formula:

> Data-Driven Decision-Making = Useful Data + Data Science + Business Acumen + Supportive Culture.

In which:

1. Useful Data = Relevant data + Quality + Structured/Modeled.
2. Data Science = Analytical Skills + Math & Statistical Skills + Technology Aptitude.
3. Business Acumen = Business Strategy + Metrics & KPIs + Process Modeling.
4. Supportive Culture = Desire for Change + Decisions Requiring Data + Continuous Learning.

This formula implies the need for investments in education, the development of a growth mindset learning culture, and fully sustaining continuous improvement efforts.

I hope these 100 lessons provide you with ideas on designing business processes and developing information solutions. As I mentioned in the introduction, in the accompanying blog http://rubiolo.net, I will share more information, including discussions on implementation and technology, as well as a contact form for you to share your questions and comments. Use them to make data-driven decisions for your organization!

References

I referenced the following resources throughout the book. They provide perspectives and details on how to think about these topics, including comprehensive guidelines and plenty of examples.

I hope you enjoy your journey in learning how to design your business processes and develop these Business Decisions Support and Information Systems, nurturing a growth mindset learning culture, through data-driven decision-making!

---

[1] Montier, J. (2010). *"The little book of behavioral investing."* Hoboken, New Jersey: John Wiley & Sons, Inc.

[2] Jim DuBois (2017). *"Six-Word Lessons to Think Like a Modern-Day CIO."* Pacelli Publishing.

[3] Carol S. Dweck (2007). *"Mindset: The New Psychology of Success."* Ballantine Books.

[4] Robert S. Kaplan & David P. Norton (1996). *"The Balanced Scorecard: Translating Strategy into Action."* Harvard Business Review Press.

[5] Gupta, P. (2004). *"Six Sigma Business Scorecard."* New York, New York, USA: McGraw-Hill.

[6] Balestracci Jr., Davis (2009). *"Data Sanity: a Quantum Leap to Unprecedented Results."* Marilee Aust.

[7] Kaplan, R. S., & Norton, D. P. (2001). *"The Strategy Focused Organization."* Harvard Business School.

[8] Kaplan, R. S., & Norton, D. P. (2004). *"Strategy Maps."* Harvard Business School.

[9] Kaplan, R. S., & Norton, D. P. (2006). *"Alignment."* Harvard Business School.

[10] Niven, P. R. (2002). *"Balanced Scorecard Step by Step."* New York, New York, USA: John Wiley & Sons Inc.

[11] Wheeler, D. J. (2000). *"Understanding Variation, 2nd Edition."* SPC Press.

[12] Wheeler, D. J. (2005). *"The Six Sigma Practitioner's Guide to Data Analysis (Vol. 6s Guide to Data Analysis)."* Knoxville, Tennessee, USA: SPC Press.

[13] Wheeler, D. J. (2003). *"Making Sense of Data."* Knoxville, Tennessee, USA: SPC Press.

[14] Wheeler, D. J. (2009). *"Twenty Things you Need to Know."* SPC Press.

[15] Savage, S. L. (2009). *"The Flaw of Averages."* Hoboken, New Jersey, USA: John Wiley & Sons Inc.

[16] Campbell, S. K. (1974). *"Flaws and Fallacies in Statistical Thinking."* Englewood Cliffs, New Jersey, USA: Prentice Hall.

[17] Darrell Huff (1954, 1982, 1993). *"How to lie with statistics."* W.W. Norton & Company, Inc.

[18] Stephen Few (2015). *"Signal: Understanding what matters in a world of noise."* Analytics Press.

[19] Wheeler, D. J. (2000). *"Beyond capability confusion, 2nd ed. – The average cost of use."* SPC Press.

[20] Wheeler, D. J. (2000). *"The process evaluation handbook."* SPC Press.

[21] Wheeler, D. J. (2010). *"Reducing production costs: how to convert Capability Indexes and performance Indexes into Excess Costs of Production and Use."* SPC Press.

[22] Lawrence Corr, Jim Stagnitto (2014). *"Agile Data Warehouse Design."* DecisionOne Press.

[23] Kimball, Ross (2013). *"The Data Warehouse Toolkit, 3rd Ed."* Wiley Publishing, Inc.

[24] Kimball, Caserta (2004). *"The Data Warehouse ETL Toolkit."* Wiley Publishing, Inc.

[25] Kimball, Ross (2016). *"The Kimball Group Reader. Remastered collection."* Wiley Publishing, Inc.

[26] Kimball, R. T. (2008). *"The Data Warehouse Lifecycle Toolkit, 2nd Ed."* Wiley Publishing, Inc.

[27] Stephen Few (2018). *"Big data, big dupe: a little book about a big bunch of nonsense."* Analytics Press.

[28] Stephen Few (2009). *"Now you see it. Simple visualization techniques for quantitative analysis."* Analytics Press.

[29] Stephen Few (2012). *"Show me the numbers. Designing tables and graphs to enlighten, 2nd ed."* Analytics Press.

[30] Cole Nussbaumer Knaflic (2015). *"Storytelling with data. A data visualization guide for business professionals."* Wiley.

[31] Naomi B. Robbins (2013). *"Creating more effective graphs. A succinct and highly readable guide to creating effective graphs."* Chart House.

[32] Harry I. Forsha (1995). *"Show me. The complete guide to storyboarding and problem solving."* ASQC Quality Press.

About the Six-Word Lessons Series

Legend has it that Ernest Hemingway was challenged to write a story using only six words. He responded with the story, "For sale: baby shoes, never worn." The story tickles the imagination. Why were the shoes never worn? The answers are left up to the reader's imagination.

This style of writing has a number of aliases: postcard fiction, flash fiction, and micro fiction. Lonnie Pacelli was introduced to this concept in 2009 by a friend, and started thinking about how this extreme brevity could apply to today's communication culture of text messages, tweets and Facebook posts. He wrote the first book, *Six-Word Lessons for Project Managers*, then started helping other authors write and publish their own books in the series.

The books all have six-word chapters with six-word lesson titles, each followed by a one-page description. They can be written by entrepreneurs who want to promote their businesses, or anyone with a message to share.

See the entire ***Six-Word Lessons Series*** at **6wordlessons.com**